CRUNCH IT, MUNCH IT

CRUNCH IT, MUNCH IT
And Other Ways to Eat Vegetables

by Shirley Parenteau / drawings by Tom Huffman

Coward, McCann & Geoghegan, Inc. / New York

To my children: David, Scott, and Cherie

Text copyright © 1978 by Shirley Parenteau
Illustrations copyright © 1978 by Thomas Huffman

Library of Congress Cataloging in Publication Data

Parenteau, Shirley.   Crunch it, munch it.

SUMMARY: A collection of 19 recipes using raw
or lightly cooked fresh vegetables. Includes
historical notes and information on vitamin and
mineral content of the foods.

1. Cookery (Vegetables)—Juvenile literature.
[1. Cookery—Vegetables]   I. Huffman, Tom.   II. Title.
TX801.P35   641.6′5   78-8582   ISBN 0-698-20466-2

Designed by Bobye List   Printed in the U.S.A.

CONTENTS

INTRODUCING VEGETABLES

A trip to the vegetable market is like traveling to many countries. You would be surprised to find only four or five vegetables in the store. But people who lived a hundred years ago would be surprised to see tomatoes (which came originally from Peru), cucumbers (from India), potatoes (from Ireland), and other vegetables from many countries all sold together.

Most of the vegetables you eat are now grown somewhere in the United States, but almost every vegetable began long ago in another part of the world. Vegetables grew as wild as the weeds we pull from our gardens, today, and often tasted tough and stringy. Modern plant scientists (called botanists) have spent many years working in laboratories and in fields to help farmers grow vegetables that are tender and full of flavor.

Some vegetables are juicy. Some are crunchy. Some (such as asparagus) are available only in certain seasons. Others (such as potatoes) are available year-round. But all vegetables are major sources of the vitamins and minerals that keep your body healthy and full of energy.

When you look at the deep orange color of a carrot, you're seeing carotene. Carotene becomes Vitamin A in your body, the vitamin needed for strong eyes, clear skin, and growth. Dark green spinach contains iron to give your body energy and lots of Vitamin A.

Red bell peppers have more than twice the Vitamin C of green bell peppers. Why? Because the red color shows that the pepper is completely ripe and filled with the vitamin that keeps your teeth and gums and bones healthy.

Many vegetables are delicious eaten raw, especially if you can pick them fresh from the garden. They are way ahead of canned or frozen vegetables in crunch or munch appeal.

Chill and rinse vegetables before you cut them. Peel root and tuber vegetables only when the skin is tough or bitter.

Vegetables can lose nutrients quickly when they are cooked, so make cooking time short and keep a lid on the pot. You want the good taste (and the vitamins and minerals) to go into you, not down the drain or up in steam. Many times, you can use a steam basket to hold vegetables above boiling water. Hot steam cooks quickly and holds in nutrients.

If you use salt, add it before eating, not before cooking. Salt draws out the tender juices.

You'll find nineteen different vegetables—in many shapes and colors and textures—in this book. You can surprise your family and friends by using the recipes (and adding your own changes) to make snacks and salads to munch on or to add a crunchy new dish to a meal. Surprise them again with the stories telling why each vegetable has been a favorite in some part of our world.

BEFORE YOU BEGIN

Ask permission to use the kitchen. Read through a recipe. Find needed ingredients and utensils and place them on your work surface. Wash your hands before preparing food.

Dry measurements should be level with the top of your measuring spoon or cup. Read liquid measurements at eye level.

Choose a mixing bowl large enough to hold all combined ingredients. Use a large, sturdy mixing spoon to stir or blend. Do not leave a metal spoon in a pan of hot liquid; it will become too hot to touch.

Ask an adult to help you at the stove until you have learned to use it safely. Remember to turn off the burner after cooking.

Use a dry pot holder with hot saucepans or skillets. Keep the handle turned toward the stove, but away from the burner. Some recipes list a steam basket inside a deep saucepan to cook vegetables above boiling water. Use a tightly fitting pan lid, but remember that steam is hot. Always open the lid away from you.

Some recipes list spatulas—long, thin blades or short, wide blades—to lift food from a pan. Others list tongs—two long, thin arms you press together to lift food from a pot or pan without cutting into it.

Use a vegetable brush with strong bristles to scrub plants you don't peel. When peeling, use a vegetable parer to shave off just a thin layer. Always pare away from your holding hand. And always cut on a board to protect your knife edge, the counter top—and your fingers.

Cut hard foods into bits by rubbing them carefully over the rough side of a grater into a bowl, and keep the grater slanted away from you. An electric blender or food processor makes grating, chopping, and blending quick and easy, but ask an adult to help you when using or cleaning the appliance.

Serve raw vegetables chilled. Place cooked vegetables on a heating tray to serve.

Metric Measurements

1 Tablespoon	15 ml (milliliter)
1 teaspoon	5 ml
1 cup	240 ml
½ cup	120 ml
⅓ cup	80 ml
¼ cup	60 ml

1. BULB

A bulb is an underground seed. Layers of stored food protect the roots, stem, leaves, and even flowers waiting to grow from deep inside.

ONION

Onion means One. It was a favorite vegetable in Egypt, where the round shape made the people think of the Universe. If they wanted you to believe what they said, ancient Egyptians wouldn't "Cross their hearts and hope to die." They would put one hand on an onion and promise to tell the truth.

Onions contain some Vitamin A, Phosphorus, and Potassium.

Lemon Cool Ones

Here's a way to make "hot" onions taste cool and sweet.

1 bunch green onions (scallions)	paring knife
lemon juice	cutting board
	tall, narrow glass or jar

Cut the roots off the green onion bulbs. Wash the stems and bulbs. Stand the onions in the glass or jar. Pour in enough lemon juice to cover the bulbs. Place them in the refrigerator overnight. The onions will be good to nibble with your lunch the next day.

5 or 6 onions

Salad or Sandwich Tip

Lemon-cooled onions are delicious in salads or sandwiches. Just take them from the jar or glass. Place them on a cutting board. Slice into thin rounds. Sprinkle them in a sandwich or scatter them over a tossed green salad.

2. ROOT

A root is usually the first part of the plant to grow from a seed. Roots take minerals and water from the soil and sometimes store food for the plant.

CARROT

Carrots were a favorite vegetable of the ancient Greeks and Romans for a very long time—for two thousand years. But carrots were new to ladies of the court in Elizabethan England. They dressed in ruffles and lace and jewels, then used the feathery green tops of carrots to decorate their hair.

Carrots have lots of Vitamin A.

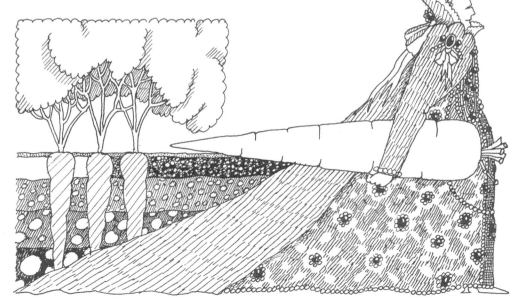

3

Sticky Snack

Make a special treat of grated carrots and celery sticks. Serve at once for the best color and nutrition.

¼ cup crunchy peanut butter	bowl
2 teaspoons honey	measuring spoons
1 medium carrot	grater or blender
3 celery stalks	measuring cup
	mixing spoon

Grate scrubbed carrot with a hand grater or blender (about ½ cup). If you use a grater, remember to slant it into a bowl. If you use a blender, ask an adult to help. First cut the carrot into ½-inch pieces. Place in blender jar. Put cover on. Turn to *Grate*. Run blender for 15 seconds.

Cut the washed celery stalks into 4-inch sticks.

Mix the peanut butter, honey, and grated carrot in a bowl with a spoon. Use the back of the spoon or a blunt knife to mound the mix into the hollow of the celery sticks and serve.

10 to 12 mounded celery sticks

BEET

Many many years ago, beets were stringy roots growing wild near the Mediterranean Sea. The seeds were carried by travelers from country to country. Today's tender beets are important in the diets of people in northern Europe. Borscht in Russia and Barszcz Czysty in Poland are names for beet soup. A winter meal in these cold countries is usually a hearty red soup made with beets and beef broth.

Beets are rich in minerals: Potassium, Magnesium, and Sodium.

Summer Borscht

Chilled borscht is great for summer and easy with a blender.

4 or 5 tender beets
2 cups water
1 beef bouillon cube
2 teaspoons lemon juice
1 teaspoon onion powder
½ cup sour cream
¾ teaspoon salt
  pepper, a few grains

vegetable brush
paring knife
measuring cup
measuring spoons
glass or plastic container
saucepan
blender

Cut the root and stem ends from the beets. Scrub under water with a brush, but don't peel. Slice the beets into ½-inch cubes (about 1½ cups). Put diced beets into a saucepan with 1 cup of the water. Bring to a boil. Cover and cook over low heat for 12 to 15 minutes. Take the saucepan off the stove and cool.

Ask an adult to help you use the blender. Pour beets and cooking water into blender jar. Cover and blend at *medium speed* for 10 seconds.

Dissolve the bouillon cube in 1 cup of boiling water. Cool. Add ½ cup of the bouillon broth to beet paste in blender. Also add all the lemon juice and onion powder, salt and pepper. Add 3 tablespoons of the sour cream. Cover and blend for 20 seconds at *high speed*. Pour into container. Stir in the other ½ cup of bouillon. Chill for 2 hours in refrigerator until ice cold. Top each dark red serving of borscht with a creamy white tablespoon of the remaining sour cream.

4 cups

3. TUBER

Tubers grow from thin underground stems. They are storage houses that fill with food for certain kinds of plants. Branches can grow from tuber "eyes" just as branches grow from stems of other plants.

POTATO

The Irish love potatoes so well, they often say, "A day without a potato is a day without food." The Irish enjoy a two-potato dish called Boxty. It's a skillet cake made from cooked mashed potatoes and grated raw potatoes. There is a legend that says when a ring is wrapped in paper and stirred into boxty batter, the finder will have a happy marriage.

Potatoes contain Potassium, some Vitamins A, B, and C.

Boxty Cakes

Boxty is good to make when you have leftover mashed potatoes.

1 cup mashed potatoes
2 large, raw potatoes
1 cup unsifted flour
2 teaspoons baking powder
2 teaspoons salt
2 eggs
¼ cup milk
　butter or margarine
¼ cup ground almonds (optional)

paring knife
blender or grater
paper towels
2 medium-size bowls
1 small-size bowl
sifter
skillet
measuring cup
measuring spoons
tablespoon
wide spatula

Bring mashed potatoes to room temperature while you peel and grate 2 large potatoes (about 1 cup). If you use a hand grater, slant it into a bowl. If you use the blender, ask an adult to help. First cut potatoes into ½-inch cubes. Put ½ cup into blender jar. Put cover on. Turn to *Grate*. Run blender for 5 seconds. Empty and repeat. Sandwich grated potatoes in 3 or 4 layers of paper towels and squeeze them over the sink to get the moisture out. Put the drained, grated potatoes into a medium-size bowl with the mashed potatoes.

　Sift the flour, baking powder, and salt into another medium-size bowl. Beat the eggs in a smaller bowl. Then stir the beaten eggs and sifted flour mixture into the bowl of potatoes. Stir in the milk, a little at a time, to make a batter. Add ground almonds, if you wish. Drop by tablespoon onto a hot, greased skillet. Remember to leave space around each spoonful. Boxty will puff into fat cakes. Cook over medium heat for about 4 minutes. When you see bubbles around the edges, use a wide spatula to turn the cakes. Brown them on the other side for another 4 minutes. Serve hot with butter.

12 3-inch cakes

4. STEM

A stem is like a river carrying salts and water from the roots through the plant to the leaves. Most stems hold the plant in the air.

ASPARAGUS

Asparagus comes from a Greek word that means "to spring up." Early in the spring, the plants push tender green tips above the soil. They're a promise of summer to come. Spring asparagus was a favorite vegetable in Rome, hundreds of years ago. When Emperor Augustus gave an order, he liked to say, "Do it quicker than you can cook asparagus."

Asparagus contains Vitamins A and C.

Speedy Spears

For the most delicate flavor, serve cooked asparagus with a light sauce.

1 pound young asparagus spears
¼ cup water
    salt or seasoned salt
¼ cup butter or margarine

deep saucepan
steam basket
measuring cup
shallow serving plate
tongs

Wash the asparagus spears. If the stems are long, bend each one until it snaps. Throw away the tough, lower part.

Pour ¼ cup of water into a deep saucepan. Place steam basket in the pan and add the tender tips. Bring the water to a boil. Cover the pan tightly. Lower the heat and steam for 8 to 10 minutes, just until asparagus is tender. Use tongs to lift the stalks carefully onto a shallow serving plate. Dust lightly with salt or seasoned salt. Serve with a side dish of melted butter to spoon over the bright green spears.

4 servings

CELERY

In northwestern Italy, near the mountains of Switzerland and France, the people like La Bagna Cauda. That means "a hot bath," but they don't climb into it. This bath is a dip for celery sticks, cabbage chunks, and green pepper slices.

Green celery has Vitamin C, Potassium, and Sodium.

Hot Bath Dip

A creamy dip with a mild taste of fish and garlic.

3 tablespoons cooking oil	small skillet
1 medium garlic clove	paring knife
¾ teaspoon anchovy paste	spoon
¼ cup heavy cream	measuring cup
3 celery stalks	measuring spoons
other raw vegetables (optional)	small bowl
Italian bread sticks	heating tray

Scrub the celery stalks and cut into 4-inch sticks. Chill them in the refrigerator with any other raw vegetables you wish to serve.

Peel the garlic and cut the clove in half.

Heat the oil in a small, heavy skillet over very low heat. Add the garlic to the oil and keep burner at low temperature for 10 minutes, stirring occasionally to get the garlic flavor. Heat gently so the garlic does not burn.

Blend the anchovy paste into the heated oil until the fish dissolves. Then add the cream, a little at a time. Stir quickly and constantly to blend the cream with the oil. When the creamy mixture is hot, take the garlic out. Pour the dip into a warmed, small bowl on a heating tray. Serve as a hot dip with crisp, chilled celery sticks (and other raw vegetables) and crunchy Italian bread sticks.

½ cup dip

SPROUTS

The people of Southeast Asia enjoy delicate tastes in food. In Thailand, cakes are flavored with the smoke of perfumed candles. In Burma, morning glory flowers are dipped into fish sauce and eaten. People of every Southeast Asian country enjoy the tangy taste of mung bean sprouts, a vegetable that is becoming a favorite in the United States.

Bean sprouts have Vitamin C and minerals.

Sprout Toss

Be sure sprouts are very fresh for the best flavor and nutrition.

¼ cup water	deep saucepan
2 cups mung bean sprouts	steam basket
2 teaspoons soy sauce	measuring cup
2 teaspoons rice vinegar or	measuring spoons
cider vinegar	small bowl
2 teaspoons salad oil	serving bowl
	tongs
	fork

Pour ¼ cup water in a deep saucepan. Place steam basket in the pan and add sprouts. Bring the water to a boil. Cover the pan tightly. Lower heat and steam for 3 minutes only. Sprouts should be cooked but still crunchy.

Mix soy sauce, vinegar, and salad oil together in a bowl. Use tongs to lift the sprouts into a serving bowl. Pour the dressing over the hot sprouts and toss with a fork until all are coated. Serve at once.

4 servings

Sandwich Tip

Leftover mung bean sprouts add a tangy taste to cheese, peanut butter, or tuna fish sandwiches.

5. LEAF

Leaves are tiny factories where water and minerals from the roots are mixed with light and carbon dioxide from the air to make food for the plant.

CABBAGE

Cabbage is such a favorite in Germany, it has become a part of the language. Instead of saying, "I've heard that story before," a German might tell you, "That's just warmed-over cabbage." If you try to talk your way out of a job, you may hear, "Pretty words do not make the cabbage fat." And if you make a dumb mistake, you'll probably be called a Kohlkopf, a "cabbage head."

Cabbage contains Vitamin A, Potassium, and Calcium.

Kohlkopf Salad

Crunchy granola cereal adds surprise to cabbage and fruit salad. Use your favorite kind.

½ medium head white or purple cabbage
1 sweet apple
½ pound seedless grapes
½ cup mayonnaise
1 cup crunchy granola cereal
lemon juice
4 lettuce leaves

4 salad bowls
grater
strainer
paring knife
measuring cup
large bowl
fork

Shred cabbage with a coarse hand grater (about 2 cups) into a bowl. Rinse unpeeled apple and grapes. Dice the apple (about 1 cup).

Line 4 salad bowls with crisp, chilled lettuce leaves. Mix the cabbage, apple, grapes, and mayonnaise together in a large bowl. Squirt with a few drops of lemon juice. Toss lightly with a fork. Divide into the bowls. Top each salad with ¼ cup granola.

4 servings

SPINACH

Back in the year 657, someone in China wrote of growing and eating spinach. It was the first written record of the vegetable, but the beginning of spinach is still a mystery. We don't know where spinach first grew as a wild plant or how it came to those Chinese gardens.

Spinach is an excellent source of Vitamin A, Potassium, and Iron.

Oriental Spinach

Water chestnuts add crunch to fresh-tasting spinach.

1 pound spinach leaves
¼ cup water
2 tablespoons salad oil
2 tablespoons soy sauce
½ teaspoon sugar
   5-ounce can of water chestnuts

deep saucepan
steam basket
measuring cup
measuring spoons
small bowl
skillet
paring knife
paper towels
fork

Wash the spinach leaves twice, lifting them from the water each time so the grains of sand stay in the sink basin. Pat the leaves dry with paper towels. Tear spinach into bite-size pieces. Put ¼ cup water into a deep saucepan. Place steam basket in pan and push in the spinach leaves. Bring the water to a boil. Cover the pan tightly. Lower heat and steam for 3 minutes.

Mix salad oil, soy sauce, and sugar in a small bowl and pour into a skillet. Drain and slice the water chestnuts. Add the steamed spinach and water chestnuts to the skillet. Warm over medium heat for 2 or 3 minutes, turning the spinach with a fork to be sure it is well coated. Serve at once for best flavor.

4 servings

LETTUCE

Lettuce is one of the first spring plants. People in ancient Greece were happy to see lettuce plants opening their leaves to the sun. This showed that winter was over. The Greeks put the young lettuce plants into pots and carried them through the streets to celebrate the return of spring.

Dark green lettuce has good Vitamin A, some Vitamins B and C, and is rich in minerals: Potassium, Calcium, and Sodium.

Walking Salad Cup

You can celebrate any season with this lettuce-cup walking salad.

1 pound berries
1 sweet apple
1 cup raisins or currants
1 juice orange
1 head lettuce

paring knife
juicer
small bowl

Cut the orange in half and squeeze out the juice. Pour into a small bowl and remove any seeds. Rinse unpeeled apple and berries. Cut the apple into small cubes and put them into the orange juice; mix until coated. This will keep the apple from turning dark. Pull leaves of lettuce away from the head, two or three for each salad. Wash the leaves and let them curl into the shape of cups. (Inside leaves have more curl.)

Put apple cubes, berries, and raisins or currants into each lettuce cup. Carry the salad in your hand to eat as a snack. Be sure to eat the crunchy lettuce cup, too.

4 salads

6. FLOWER

Bright blossoms look pretty and smell nice for a good reason. Their job is to attract insects to carry pollen from one plant to another so seeds can grow to make new plants.

CAULIFLOWER

Arabians first brought cauliflower to Spain. Then it appeared in other countries. Today, if you joined a formal dinner in a Middle Eastern country (Egypt, Israel, Lebanon, Turkey), you might first be served Meza, a feast of appetizers. With a choice of up to fifty different foods, you could munch on olives, cubes of goat cheese, broiled baby octopus, round combs of honey, and sprigs of cauliflower to dip into a tangy sauce.

Cauliflower gives you some Vitamins A, B, and C, and minerals.

Creamy Cauliflower Dip

Add a tangy taste to fresh cauliflower sprigs.

1 head cauliflower
1 package dry salad dressing mix
1 tablespoon lemon juice
1 tablespoon water
⅓ cup sour cream or
   plain yogurt

measuring spoons
measuring cup
small bowl
serving bowl

Discard the outer leaves of the cauliflower head. Break the vegetable into bite-size flowerets and wash well.

Stir the lemon juice and water into the salad dressing mix in a small bowl. Blend in the sour cream or yogurt. Cover and chill in the refrigerator. Serve the flowerets with the creamy dip.

½ cup

BROCCOLI

Broccoli is one of the oldest vegetables in the cabbage family, but it wasn't known in the United States until the 1920's, when an Italian vegetable grower moved to California. He advertised the new vegetable—broccoli—over a new invention—radio. Soon, fresh shipments of broccoli traveled by train from the West to hungry buyers in the East.

Broccoli is rich in Vitamins A, B, and C, and in Potassium.

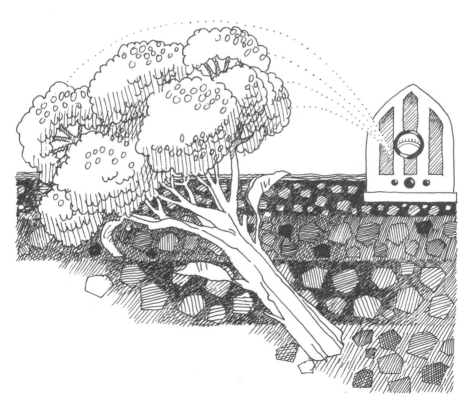

Trees and Cheese

A crunchy, cheesy way to serve broccoli.

1 head broccoli	paring knife
4 tablespoons butter or margarine	skillet
salt	tongs
pepper	measuring cup
¼ cup grated Parmesan cheese	measuring spoons

Tear away the heavy outer leaves of the broccoli head and cut off the lower 2 inches of the thick stalks and discard. Cut the shortened stems into ½-inch slices. Separate the broccoli top into flowerets of little green trees. Wash well.

Melt the butter or margarine in a heavy skillet over medium heat. Add the stem slices. Heat for 5 minutes, turning often with tongs so all pieces are buttered and warmed through. Add flowerets to the skillet. Heat for 5 minutes more, turning occasionally. Sprinkle lightly with salt and pepper and grated Parmesan cheese. Serve at once.

4 servings

7. SEED

Seeds are the most important parts of the plant. All the other parts grow to form seeds so new plants can grow to replace them.

BEANS

Kidney, lima, black, green, and wax—beans come in many shapes and colors. Some are eaten green with their pods. Others are dried and cooked in soups or stews. Beans are among our most ancient vegetables and appear in stories from many lands. Some beans (soya) contain as much protein as steak. Maybe that's why, in Italy, people say, "The more beans you eat on New Year's Day, the more wealth you'll have in the coming year."

Green beans contain Vitamin A and Potassium.

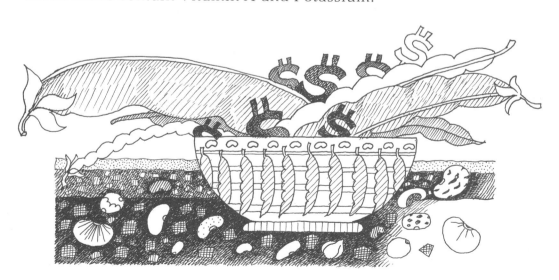

Green Bean Fondue

Crisply cooked green beans dipped into spicy fondue.

1 pound green beans
¼ cup water
8-ounce can tomato sauce
¼ teaspoon dried oregano
⅛ teaspoon dried basil
1 cup grated cheddar cheese

deep saucepan
steam basket
small skillet
measuring cup
measuring spoons
grater
spoon
warm bowl
heating tray
fondue forks

Wash the beans. Snap off stems and break into pieces 2 inches long. Put ¼ cup water in a deep saucepan. Place steam basket in the pan and add beans. Bring water to a boil. Cover tightly. Lower heat and steam for 5 minutes.

Pour tomato sauce into a small skillet. Crumble herbs and stir into sauce. Keep on low heat. Add grated cheese, a little at a time, stirring until it melts. When heated through, pour into a warmed bowl on a heating tray. Spear the beans with a fork and dip into the fondue.

4 servings

PEAS

When peas were first grown in France, in the late seventeenth century, people couldn't get enough of the new vegetable. They would go to a party and eat peas until they were full, then go home and eat more before going to bed. One French woman wrote to her friend, "Peas are both a fashion and a madness."

Peas contain lots of Vitamin A, some Vitamins B and C, and are rich in minerals: Iron, Magnesium, Potassium, Phosphorus.

Peppermint Pods

Pods and mint give peas extra flavor.

1 pound tender young peas
¼ cup water
2 teaspoons fresh mint leaves or
   ¼ teaspoon dried mint flakes
½ teaspoon salt
2 tablespoons butter or margarine

saucepan
measuring spoons

Pop the peas from their pods. Wash and save 3 pods. Put ¼ cup water in a saucepan and heat to a boil. Add the peas and the washed pods. Cover with a tight lid and cook over low heat for 8 minutes or just until tender. Remove the pods from the pan. Stir in the fresh chopped (or dried) mint, salt, and butter. Serve at once.

4 servings

CORN

Columbus found Indians growing corn when he arrived in the New World, but their corn was a garden plant that needed care. No one knew how corn first began to grow. A few years ago, scientists found seeds four thousand years old in a cave in New Mexico. The seeds were from a corn-like grass. Botanists believe our corn changed over many years from wild grass growing in Mexico and Central America.

Corn is rich in Potassium and Phosphorus, and has Vitamin A.

Roasted Cobs

Oven-roasted corn kernels are chewy, with extra flavor from the husk.

fresh young ears, with husks	roasting pan
softened butter or margarine	cotton twine
slice of bread, folded	pot holders
salt	

Pull back the corn husks, one at a time, opening the ear like the petals of a flower. Keep the husks fastened at the stem end. Take out the soft corn silks. Spread pats of butter or margarine over the ears, using a folded slice of bread as a spreader. Pull the husks back in place over the buttered ears.

Set the oven to 450° F. Use cotton twine around the open end to tie the husks snugly together. Sprinkle the outside of the ears with water so they are very damp and will not burn. Place them in a roasting pan and put them in a hot oven for 15 minutes. Use pot holders to remove the pan from the oven. Serve with salt.

Picnic Tip

Prepare corn husks in the same way to roast cobs in picnic or campfire coals. After the cobs are tied together and dampened, wrap each husk-covered ear with 2 layers of heavy foil. Use long-handled barbecue tongs to place the ears on the grill or on the edge of the coals. Roast the cobs for 20 to 25 minutes, turning occasionally with the tongs. Remember, the steam inside the foil is hot. Ask an adult to help you open the wrap. Use the foil as a picnic plate.

8. SEED HOLDER

The part of the plant that encloses the seed is called the fruit, whether it's an apple or a cucumber.

TOMATO

A court of law was once asked to decide whether tomatoes were fruit or vegetable. Vegetable won because we eat tomatoes usually in a salad with the main part of our meal, instead of as a dessert. For many years, people believed tomatoes were poisonous. Later they were called love apples and given as presents to show affection. Wild tomatoes still grow in Peru, where they climb over trees and bushes on vines that can grow fifty feet long.

Tomatoes are rich in Potassium and Vitamin A, and have Vitamins B and C.

Herbed Cherry Tomatoes

Whole cherry tomatoes for a bright and tangy salad.

12 plump cherry tomatoes
2 green onions
⅔ cup salad oil
¼ cup vinegar
½ teaspoon dried basil or
   herb mix
1 teaspoon salt
¼ teaspoon pepper
   lettuce leaves

jar with lid
measuring cup
measuring spoons
knife
cutting board
salad bowl

Blend the oil, vinegar, basil or mixed herbs, salt, and pepper in a jar or container with a lid. Slice the green onions into thin rounds. Put the tomatoes and the sliced onions into the jar and fit the lid on tightly. Shake gently so the tomatoes are well covered with dressing. Chill in the refrigerator for several hours or overnight.

   The next day shake gently, then scoop the tomatoes out and arrange them on lettuce leaves in a salad bowl. Pour the dressing over them and serve.

4 servings

Salad Tip

The herbed tomatoes and dressing are good when added to a regular tossed green salad (made without dressing) the following day.

CUCUMBER

If you visit a September festival in India, you may see parades of elephants painted with flowers. During a March festival, colored dye powders may be thrown over your clothes like paper confetti. At any time, you would be invited to taste native foods, including one you can find in your market every day. It's the cool cucumber. This vegetable grew first in deep valleys below the snowy Himalaya Mountains.

Cucumbers have some Vitamin C, and are rich in minerals, especially Phosphorus, Sodium, and Calcium.

Confetti Cucumber

A colorful, flavorful way to serve chilled cucumbers.

1 medium-size cucumber
  3-ounce package of cream cheese
⅛ teaspoon chili powder, paprika *or*
  curry powder
  lettuce leaves
2 tablespoons creamy salad dressing

vegetable parer
paring knife
fork
spatula
measuring spoons
plastic wrap
paper towel
salad plate

Peel the cucumber skin with a vegetable parer. Cut off ends. Run fork tines down the length of the cucumber to make grooves. Repeat on all sides, then slice in half. Scoop out the seeds. Pat the hollows dry with a paper towel.

Stir the softened cheese with a fork until creamy. Add the spice powder. Use a spatula to spread the cheese mixture smoothly inside the cucumber halves. Press the halves back together. Roll tightly in plastic wrap. Chill in refrigerator for 1 hour or longer. Place lettuce leaves on a salad plate. Cut cucumber into ½-inch slices and arrange them on lettuce bed. Pour creamy salad dressing over them before serving.

3 servings

BELL PEPPER

Columbus made more than one mistake. He was looking for spices when he sailed to what he called the East Indies. When he saw sweet peppers growing in what we know as the West Indies, he thought they were the same plant that grew the black peppercorns people grind and sprinkle over food. He named the plants Pepper and the name stayed. Today we say bell pepper or sweet pepper when we mean the vegetable and not the spice.

Peppers are rich in Vitamins A and C, and have some Vitamin B and minerals.

Relish Bells

Sweet peppers make pretty relish cups you can eat.

2 red *or* green bell peppers
  lettuce leaves
  seasoned salt (optional)
  carrot sticks, radishes,
    cauliflower buds

paring knife
cutting board
relish tray

Use a paring knife to cut around the stem end, then cut each pepper in half to make two cups. (One will have no bottom.) You can make the cups fancier by cutting diagonal slits or V's at the top, like teeth all the way around. Pull the halves apart. Pull out the stem and seeds and scrape away the white membrane along the sides.

Arrange lettuce leaves on a relish tray. Place pepper cups on lettuce. Sprinkle with seasoned salt, if you wish. Fill with cut-up raw vegetables. Eat the relishes—then eat the cups.

4 relish cups

Dip Tips

Serve Relish Bells with Creamy Cauliflower Dip (page 22) for a bright color contrast. Add other raw vegetables. Then slice the pepper cups into sticks to dip and eat, too.

Fill Relish Bells with celery sticks to dip into the Hot Bath Dip (page 12). Then slice the pepper cups to dip and eat, too.

SQUASH

The native Americans dried small wild squash to make the seeds rattle inside for dance music. When they learned to grow the vegetables bigger for better rattles, the Indians also grew squash that was more tender and better to eat. Squash comes from As-kutasquash, which means "Green thing eaten green" in the Algonquian language. All squash belong to a large family that also includes pumpkins, gourds, cucumbers, cantaloupes, and watermelons.

Squash is rich in minerals, especially Potassium, in Vitamin B, and has Vitamins A and C.

Acorn Rings

Flower-like acorn rings are delicious with chilly applesauce centers.

2½ cups water
1 acorn squash
1 large bay leaf
½ cup applesauce
salt
pepper

deep saucepan
skillet with cover
large knife
wide spatula
measuring cup
serving dish
tablespoon

Chill applesauce in the refrigerator while preparing the squash. Acorn squash are very hard-shelled. Make them easy to slice by first parboiling. (Pour 2 cups of the water into a deep saucepan. Put in the squash. Bring water to a boil. Cover tightly. Lower the heat and boil gently for 30 minutes.) The heavy squash can slip and splash hot water, so ask an adult to remove it from the pan.

When cool, cut the squash into 1-inch rings. Put ½ cup water and the bay leaf in a heavy skillet. Place squash rings in skillet. Bring water to a boil. Cover. Lower heat and simmer for 20 minutes. Remove rings carefully on a wide spatula. Place each ring on a separate serving dish. Sprinkle lightly with salt and pepper. Spoon 2 tablespoons of chilled applesauce in the center of each ring and serve.

4 servings